THE SCIENCE BEHIND STUFF

Angela Royston

The medical procedures described in this book were carried out in the past and should not be attempted today.

Published 2011
First published in hardback 2010 by
A&C Black Publishers Ltd.
36 Soho Square, London, W1D 3QY

www.acblack.com

ISBN 978-1-4081-2906-7

Series consultant: Gill Matthews

Text copyright © 2010 Angela Royston

The right of Angela Royston to be identified as the author of this work has been asserted by her in accordance with the Copyrights, Designs and Patents Act 1988.

A CIP catalogue for this book is available from the British Library.

Every effort has been made to trace copyright holders and to obtain their permission for use of copyright material. The authors and publishers would be pleased to rectify any error or omission in future editions.

This book is produced using paper that is made from wood grown in managed, sustainable forests. It is natural, renewable and recyclable. The logging and manufacturing processes conform to the environmental regulations of the country of origin.

Produced for A&C Black by Calcium. www.calciumcreative.co.uk

Printed and bound in China by C&C Offset Printing Co.

All the internet addresses given in this book were correct at the time of going to press. The author and publishers regret any inconvenience caused if addresses have changed or sites have ceased to exist, but can accept no responsibility for any such changes.

Acknowledgements

The publishers would like to thank the following for their kind permission to reproduce their photographs:

Cover: Shutterstock. **Pages:** Bayer Business Services: Corporate History & Archives 22t, 22b; Corbis: Bettmann 27; Courtesy of the National Library of Medicine 7b, 8, 13, 14, 18, 19, 21, 23; Getty Images: Hulton Archive 10; Library of Congress: 9t, 20t; National Archives: 25b; Photolibrary: 24; Public Health Image Library: CDC/Jean Roy 9b, CDC/Don Stalons 25t; Shutterstock: Oguz Aral 6, Linda Bucklin 5t, Condor36 29, Daisy Daisy 26, Christian Darkin 15b, Jubal Harshaw 15t, Girish Menon 5b, Dario Sabljak 20b, Michael Taylor 17; Wikimedia Commons: 7t, 11, 16, Ernest Board 12, Rick Proser 28, Peter Treveris 4.

Contents

Good News

▼ Trepanning was believed to relieve pressure on the brain, but it was quite gruesome!

People are excited when a scientist discovers a new treatment for an illness or an injury. A medical breakthrough can lead to better ways of curing people.

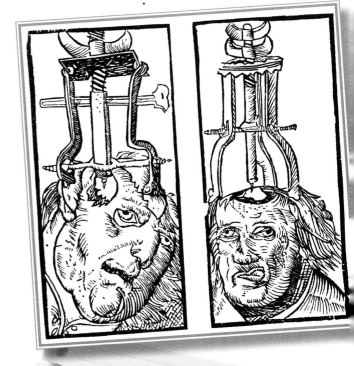

DEADLY DISEASES

Four hundred years ago, doctors had few ways of helping sick people. People died from illnesses, such as scarlet fever, that are easily cured today. Doctors did not understand how people caught diseases and so they could not prevent them.

BORING CURE – BAD IDEA

People have been trying to cure disease since prehistoric times. Ancient skulls have been found with holes **bored** through them. The holes were bored while the person was still alive – this is called trepanning. It was done to let out the evil spirits, which, people believed, caused illness and pain.

FATAL WOUNDS

Injuries, such as bad cuts and broken bones, were often fatal, because the wounds became infected. Doctors did not understand what caused **infection** and could not cure it.

TESTING CURES

Some cures did more harm than good, until scientists and doctors realized that they had to test their ideas and cures to see if they worked. Today, new medicines have to be tested for several years before they are used on sick people.

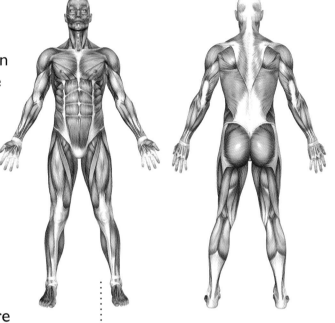

▲ Observing the inside of the human body helped surgeons to understand how it works.

▼ Surgeons operate on a person in hospital. They now use machines and techniques that were once medical breakthroughs.

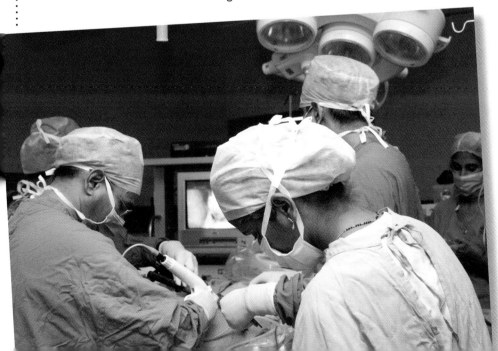

5

The Heart is Just a Pump

1628

Dr William Harvey, doctor to King Charles I, has just published an astonishing idea. He says that the heart pumps blood around the body.

▲ The heart is two pumps. The left side pumps blood to the lungs. The right side then pumps the blood around the body.

A REVOLUTIONARY IDEA!

According to Dr Harvey, blood leaves the heart through the **arteries**, and returns through the **veins**. "The movement of blood occurs constantly in a circular manner," he claims.

PRECIOUS HEART – BAD IDEA

The Egyptians thought that their heart held their soul and was where their thoughts and feelings came from. When someone died the body, and heart were mummified, but the brain was thrown away.

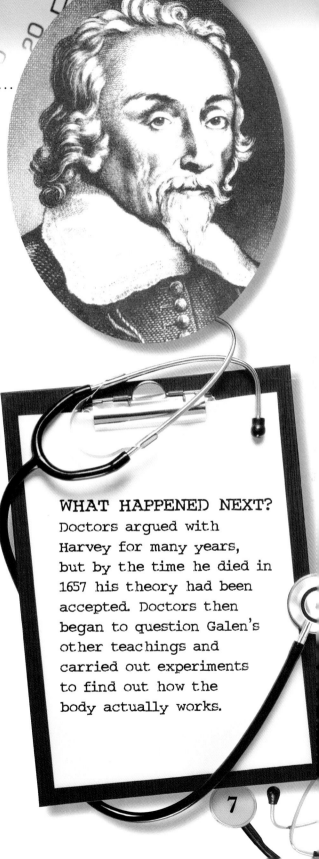

Dr William Harvey ▶

DOCTORS SAY THAT HARVEY IS MAD

"He is crack-brained," says Jean Riolan, France's leading doctor. Like other doctors, Riolan follows the ancient teachings of Galen. He was doctor to **gladiators** and four Roman **emperors** 1,500 years ago.

▼ This drawing by Dr Harvey shows that arteries and veins have valves that allow blood to flow in one direction only.

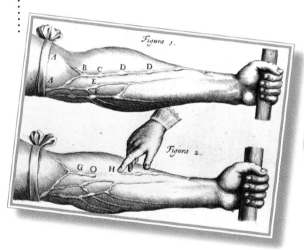

WHO IS RIGHT?

Galen said that blood is made in the liver. As it travels through the body, it is used up. But Dr Harvey insists that his experiments on dead animals, such as deer, prove that blood moves continuously around the body and is not used up.

WHAT HAPPENED NEXT?

Doctors argued with Harvey for many years, but by the time he died in 1657 his theory had been accepted. Doctors then began to question Galen's other teachings and carried out experiments to find out how the body actually works.

No More Pock Marks

Dr Edward Jenner has found a way to protect people from smallpox. He calls it inoculation. Is this the end of the terrible disease?

▲ Dr Edward Jenner

COWPOX

Every dairymaid knows that she will never catch smallpox if she has already had cowpox. Cowpox is similar to smallpox, but milder, and it can pass from cows to people. Jenner developed his inoculation using pus from cowpox blisters.

HORRORS OF SMALLPOX

In AD 400, an Indian medical book described a disease in which "the pustules are red, yellow, and white and they are accompanied by burning pain". The disease was probably smallpox. Smallpox killed about 80 per cent of children who caught it.

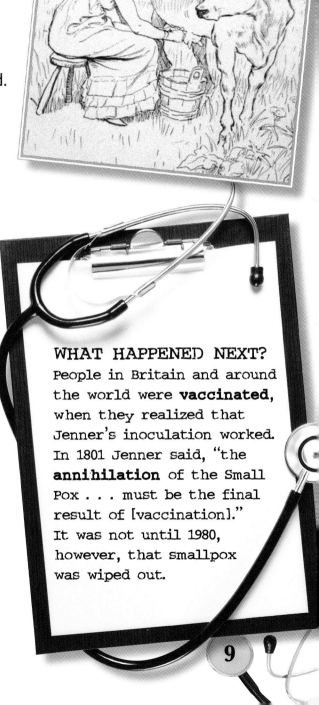

Dairymaids do not have **pock marks**. ▶
They have clear skin, because they
do not catch smallpox.

THE FIRST INJECTION

James Phipps, an eight-year-old boy, is
one of the first people to be inoculated.
On 14 May 1796, Jenner scratched
James's arms with pus from a cowpox
blister. Then on 1 July, Jenner infected
James with pus from a smallpox blister.
James did not catch smallpox – the
inoculation worked!

▼ This boy's skin is covered in smallpox
blisters. One day, smallpox may be
wiped out thanks to inoculation.

WHAT HAPPENED NEXT?
People in Britain and around
the world were **vaccinated**,
when they realized that
Jenner's inoculation worked.
In 1801 Jenner said, "the
annihilation of the Small
Pox . . . must be the final
result of [vaccination]."
It was not until 1980,
however, that smallpox
was wiped out.

Embarrassed Doctor Invents Stethoscope

1816

French doctor René Laennec has invented an instrument for listening to a patient's heart. It is called a stethoscope and he has invented it to avoid embarrassment!

REVEALING SOUNDS

William Harvey (see page 6) said that the heart should sound like "two clacks of a water bellows". Doctors also listen to the patient's breathing. They can hear if there is liquid in the breathing tubes.

TOO CLOSE FOR COMFORT

For centuries, doctors have listened to the heart beating by putting their ear to the patient's chest. Laennec could not bring himself to do this to a young woman patient. Instead, he grabbed several sheets of paper and rolled them up. He put one end of the roll of paper on the woman's chest and his ear to the other end. He knew the sound of her heartbeat would pass along the rolled tube, but he was amazed to hear her heart even louder and clearer than usual!

▼ Dr Laennec uses his stethoscope to listen to a boy's lungs.

WHAT HAPPENED NEXT?

Laennec made a proper stethoscope from wood, and used it to study diseases such as pneumonia and tuberculosis. The stethoscope has since become one of the most important instruments for doctors. It led them to **diagnose** diseases by examining their symptoms.

We Have Conquered Pain!

October 1846

The first painless operation took place at Massachusetts General Hospital yesterday. William Morton used anaesthetics – a gas that saves the patient from the agonizing pain of a major operation.

▼ William Morton anaethetizes his patient.

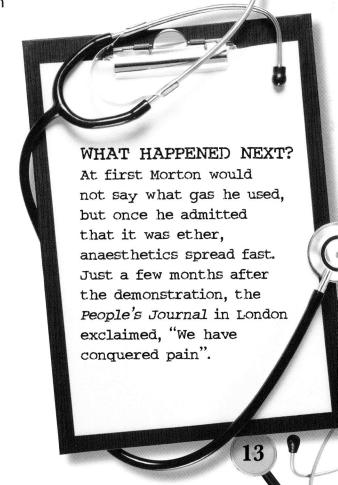

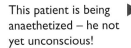

This patient is being anaethetized – he not yet unconscious! ▶

EXTRAORDINARY DEMONSTRATION

Before surgeon John Collins Warren removed a **tumour** from Gilbert Abbott's neck, Morton gave Abbott a gas to breathe in. The gas kept him **unconscious** throughout the operation. When he recovered, Abbott claimed that he had felt nothing, not a jot of pain. Doctors and medical students watched Warren carry out this extraordinary event.

OPERATING WITHOUT ANAESTHETICS

Without Morton's anaesthetizing gas, patients scream in agony during operations. Some are punched unconscious, while others are given alcohol or **opium**, but nothing fully deadens the pain. Most patients are simply held down while the operation is carried out.

WHAT HAPPENED NEXT?
At first Morton would not say what gas he used, but once he admitted that it was ether, anaesthetics spread fast. Just a few months after the demonstration, the *People's Journal* in London exclaimed, "We have conquered pain".

How Life Begins

1858

Nearly twenty years ago this newspaper reported the amazing fact that every living thing is made up of tiny cells. Now German scientist Rudolf Virchow has discovered that every cell comes from another living cell!

▲ Rudolf Virchow

DISCOVERY OF CELLS

In 1838, two German scientists, Theodor Schwann and Matthias Schleiden, realized that all living things, both plants and animals, are made up of tiny cells. Schwann thought that cells formed from a non-living substance, which he called "blastema". He was wrong!

LOGICAL CONCLUSION

In 1880, scientist August Weissman pointed out that, if all cells come from pre-existing cells, then all cells "can trace their ancestry back to ancient times". In other words, every living thing could be descended from one original cell!

This slice through the root of a fern has been magnified to show that it is made of cells. ▶

THE NEW THEORY

Virchow's new book describes how existing cells divide to form new cells. "Every cell comes from another living cell," he says. Even more amazing is his claim that every individual grows from a single cell. This first cell divides over and over again to form all the cells in the body.

▼ An amoeba is a tiny animal made up of just one animal cell. Amoebas are blobs that can easily change shape.

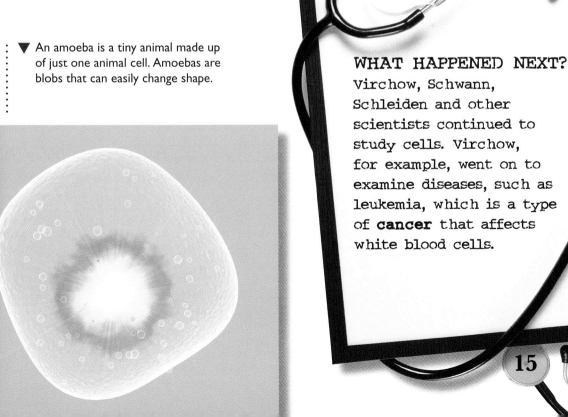

WHAT HAPPENED NEXT?
Virchow, Schwann, Schleiden and other scientists continued to study cells. Virchow, for example, went on to examine diseases, such as leukemia, which is a type of **cancer** that affects white blood cells.

15

Germs All Around Us

Finally, Louis Pasteur has proved that **germs** exist! This great French scientist has argued for years that germs exist and are all around us, but other scientists insulted and mocked him – until yesterday.

....

Louis Pasteur at work in his laboratory. ▶

16

WINNING THE ARGUMENT

In 1854, Pasteur was asked to find out why some beer and wine turns sour and has to be thrown away. But when he announced in 1857 that tiny living things, called microbes or germs, do the damage, no one believed him. Yesterday, however, Pasteur invited a group of famous scientists to the University of Paris. He proved to them beyond doubt that germs exist and they can be killed by heat.

WHAT HAPPENED NEXT?

Louis Pasteur believed that some diseases, such as cholera and typhoid, were caused by germs. Following Jenner (see page 8), he tried to create vaccines against these diseases. By 1885 he had a vaccine against rabies. Today there are vaccines against most deadly diseases.

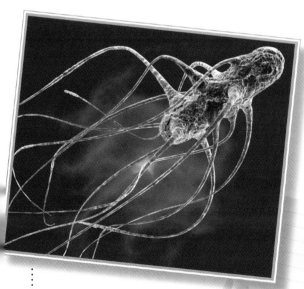

▲ A single salmonella bacteria is revealed under a microscope. Salmonella is a form of food poisoning.

PRESERVING MILK

Untreated milk soon turns sour. Pasteur showed that, if the milk is heated, the germs are killed and the milk stays good. This process came to be known as pasteurization.

Carbolic Acid Kills Germs

1867

Surgeon Joseph Lister has saved a 13-year-old's badly wounded arm by washing it with carbolic acid! The arm has now healed. Lister claims this is because the dangerous acid killed the germs in the wound.

▲ Joseph Lister

DEADLY ROT

Hospital wards that do not use antiseptics smell of patients' rotting wounds. Without antiseptics, nearly half of all patients who have a limb amputated die, because the wound becomes infected with germs.

THE ACCIDENT

When the boy was brought into Glasgow Royal Infirmary, his arm was badly cut and the bones were broken. Lister says that he would normally have amputated the arm at the shoulder. Because he knew how to use **antiseptic** carbolic acid, however, he did not hesitate to try to save the limb.

▼ Lister has used carbolic acid during operations since 1865. It healed wounds which otherwise would have become infected.

WHAT HAPPENED NEXT?

As well as using carbolic acid during surgery, Lister made his surgeons wash their hands and spray their instruments with carbolic acid. His methods were so successful, other hospitals also used them and surgery became much safer.

X-ray – the Invisible Ray

William Conrad Roentgen has discovered an invisible ray. The ray is called an X-ray and it can photograph bones – astounding!

This X-ray shows ▶
a skull and the
bones at the top
of the spine.

▲ German scientist
William Conrad
Roentgen

DISCOVERED BY ACCIDENT

Roentgen discovered the rays by accident. On 8 November 1895, he was experimenting with a **cathode ray** tube that had been tightly covered with black cardboard. The room was dark and he noticed that, whenever he passed electricity through the tube, a piece of photographic paper lying nearby glowed. He concluded that invisible rays were passing from the tube through the cardboard and making the **chemical** on the paper glow.

METAL AND BONES

Roentgen investigated the rays for several weeks. He found that they could pass through most things, even a thick book. On 22 December, he passed the rays through his wife's hand. They created a shadow of the bones in her hand and her two rings, because only the bones and the metal had blocked the rays.

WHAT HAPPENED NEXT? Roentgen published his discovery in a German scientific journal on 5 January 1896 and the news spread fast around the world. Today, X-ray machines are used by doctors and dentists everywhere to examine bones and teeth.

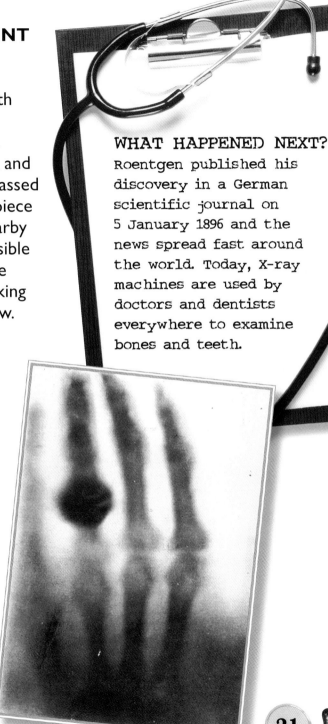

The first X-ray ever shows ▶ Roentgen's wife's hand.

Wonder Drug Discovered

▼ Felix Hoffmann

1899

A wonderful new medicine has just been launched. It is called Aspirin and it relieves severe aches and pains!

SURPRISE DISCOVERY

The drug was made by Felix Hoffmann, a German chemist. He added a group of chemicals called acetyl to salicylic acid. Salicylic acid is known to reduce pain, but Hoffman was surprised to find that the new medicine reduces swelling, too. It also lowers the high temperature of patients with **fever** Hoffman's father is one person who will benefit from the new medicine. He suffers severe pain due to **rheumatism**.

Powdered Aspirin is ▶ now available in bottles!

CENTURIES-OLD REMEDY

Some plants contain natural salicin. For centuries they have been ground up and taken to relieve pain. However, salicin irritates the stomach and so the pure acid is only given to people who, like Hoffman's father, are in extreme pain.

▲ Laudanum is now available for babies too!

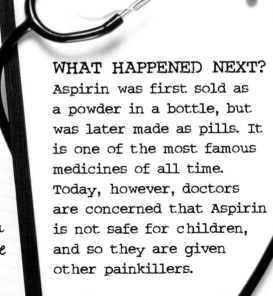

LAUDANUM—BAD IDEA

Laudanum is a form of opium and is now illegal (against the law). Before Aspirin, however, many people took laudanum to deaden toothache and other pains. Mothers even gave it to babies to stop them crying when they were teething.

WHAT HAPPENED NEXT?

Aspirin was first sold as a powder in a bottle, but was later made as pills. It is one of the most famous medicines of all time. Today, however, doctors are concerned that Aspirin is not safe for children, and so they are given other painkillers.

23

Miracle Cure

The Nobel Prize for Medicine has been awarded to these three great scientists – Alexander Fleming, Howard Florey, and Ernst Chain – for the discovery and development of **penicillin**.

▲ Fleming, Florey, and Chain being presented with the Nobel Prize for Medicine.

BACTERIA KILLER

"When I woke up on 28 September 1928," said Alexander Fleming, "I certainly didn't plan to revolutionize medicine by discovering the world's first antibiotic, or bacteria killer." But that is exactly what he did.

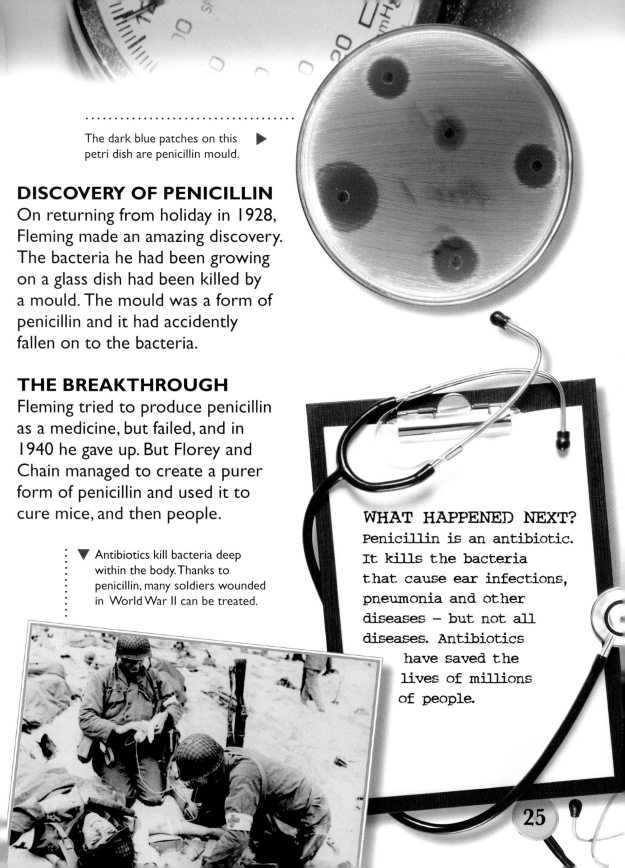

The dark blue patches on this petri dish are penicillin mould. ▶

DISCOVERY OF PENICILLIN

On returning from holiday in 1928, Fleming made an amazing discovery. The bacteria he had been growing on a glass dish had been killed by a mould. The mould was a form of penicillin and it had accidently fallen on to the bacteria.

THE BREAKTHROUGH

Fleming tried to produce penicillin as a medicine, but failed, and in 1940 he gave up. But Florey and Chain managed to create a purer form of penicillin and used it to cure mice, and then people.

▼ Antibiotics kill bacteria deep within the body. Thanks to penicillin, many soldiers wounded in World War II can be treated.

WHAT HAPPENED NEXT?
Penicillin is an antibiotic. It kills the bacteria that cause ear infections, pneumonia and other diseases — but not all diseases. Antibiotics have saved the lives of millions of people.

25

MRI Scanner Sees Everything

1977

An amazing new machine, called an **MRI** scanner, will allow doctors to see everything inside the body. Unlike an **X-ray**, which shows only hard bones, an **MRI** scan shows all the soft parts as well. It even shows cancer tumours.

▼ An MRI scan of a person's brain

INDOMITABLE

American doctor Raymond Damadian has created a machine that can scan humans for diseases such as cancer. In 1970, Damadian first noticed that a machine that combines **radio waves** and **magnetism** could detect cancer tumours in rats. Other scientists did not think it would work, but Damadian has spent the last seven years building his own scanner. He calls it *Indomitable*.

▲ Damadian explains how his MRI scanner works.

THE FIRST SCAN

On 3 July 1977 Damadian and his team carried out the first human body scan. It took 4 hours and 45 minutes, but it produced a clear picture of the heart, lungs, and chest. It has convinced his colleagues that the machine will work.

STRANGE EXPERIENCE

Having a scan can feel strange. The patient lies on a table, which then moves into the scanner's narrow tunnel. The tunnel is hot and the machine is very noisy. Some scans take 30 to 40 minutes, but others are done in seconds.

WHAT HAPPENED NEXT?

MRI scanners now work faster and have been used in hospitals around the world since the 1980s. They are used to diagnose many kinds of problems from strained muscles to tumours in the brain.

27

New Hearts in the Future

In France, Professor Alain Carpentier has invented an artificial heart. "I couldn't stand seeing young, active people aged 40 dying from massive heart attacks," he said. He has spent nearly 20 years developing the artificial heart, and expects it to be ready to use in patients by 2011.

NEW HEARTS NEEDED
Heart failure kills up to 17 million people worldwide every year. These people desperately need new hearts. Different groups of scientists are working hard to produce entirely new hearts.

▲ One of the first artificial hearts ever made

28

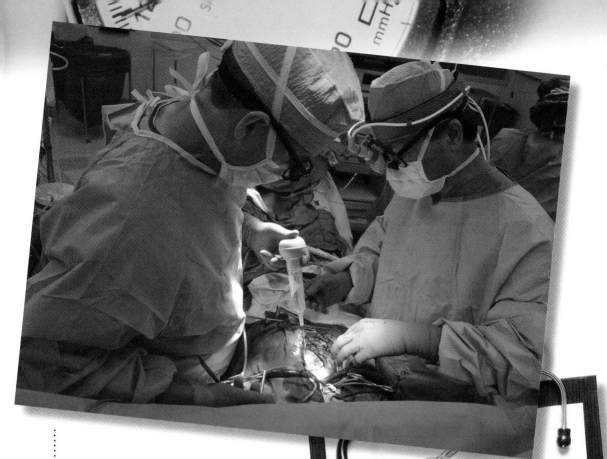

▲ Surgeons operate on a patient's heart to make it work better. When a heart is very damaged it is difficult to keep it working.

HEART TRANSPLANTS

In 1967, South African doctor Christiaan Barnard carried out the first successful **heart transplant** on patient Louis Washkansky. Since then, heart transplants have become common, but there are not enough hearts for everyone who needs them. This is why scientists are developing other solutions.

WHAT WILL HAPPEN NEXT?

Scientists in Australia have already grown new healthy, heart muscle from a small piece of existing heart muscle. Meanwhile scientists in the United States have grown new skin and new bone, using existing skin and bone.

Glossary

annihilation complete destruction

antiseptic substance that kills germs

arteries tubes that carry blood away from the heart

bored made a hole

cancer disease in which cells in the body grow out of control

cathode ray stream of electrically charged particles

cells smallest building blocks of living things

chemical powerful substance found in both natural and man-made things

diagnose to identify a disease

emperors men who rule an empire

fever very high body temperature

germs tiny living things that can cause disease

gladiators people who fought in public arenas during Roman times

heart transplant to transfer a living heart from one body to another

infection disease caused by germs

inoculation making a person able to resist a disease by infecting them with mild or dead germs

magnetism forces of attraction and repulsion made by magnets

opium powerful drug

penicillin drug made from a mould that can kill bacteria

pock marks scars left on the skin after blisters have healed

radio waves form of energy produced by combining electricity and magnetism

ray beam of energy

rheumatism painful disease that affects the joints

tumour growth or swelling

unconscious not aware of what is happening

vaccinated to be given a vaccine in order to protect against disease

veins tubes that carry blood back to the heart

Further Information

WEBSITES

Find out more about Edward Jenner and the part he played
in the history of smallpox at:
www.bbc.co.uk/history/british/empire_seapower/smallpox_01.shtml

Find out how William Morton brought anaesthetics to surgery at:
neurosurgery.mgh.harvard.edu/History/gift.htm

Learn how Louis Pasteur struggled to get his discovery accepted
by other scientists at:
www.historylearningsite.co.uk/louis_pasteur.htm

Discover the full story of how Joseph Lister used carbolic acid to kill
germs and make surgery safer at:
**www.surgical-tutor.org.uk/default-home.htm?surgeons/lister.
htm~right**

BOOKS

Graphic Discoveries Medical Breakthroughs by Gary Jeffrey.
Franklin Watts (2009).

The 10 Most Significant Medical Breakthroughs by Denis Carr.
Franklin Watts (2008).

Index